INSECTS

9780603005237

D1810433

BY

GEORGE S. FICHTER

PRINTED IN
DEAN & **SON Ltd.**
41/43 Ludgate Hill
GREAT BRITAIN
LONDON EC4
TRADE MARK

CONTENTS

603 00523 3

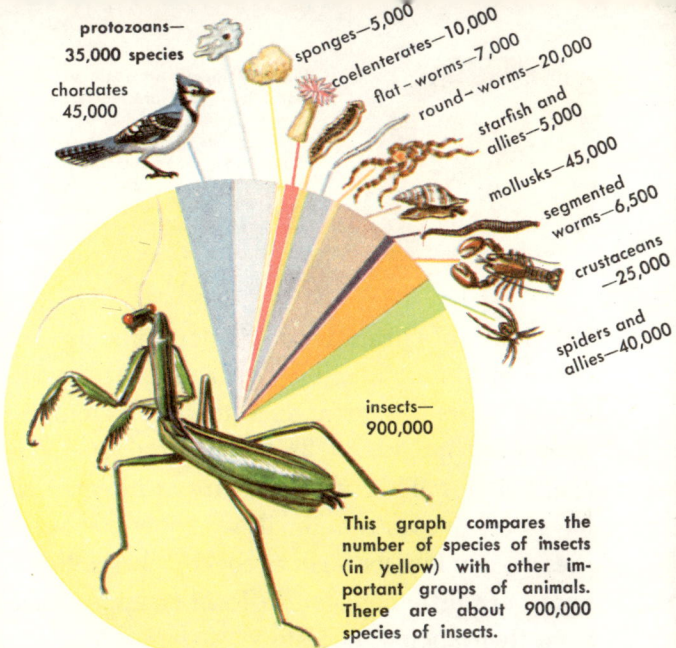

protozoans—
35,000 species

sponges—5,000

coelenterates—10,000

flat-worms—7,000

round-worms—20,000

chordates
45,000

starfish and
allies—5,000

mollusks—45,000

segmented
worms—6,500

crustaceans
—25,000

spiders and
allies—40,000

insects—
900,000

This graph compares the number of species of insects (in yellow) with other important groups of animals. There are about 900,000 species of insects.

If all the insects in the world were put in one pile and all the other land animals in another, the pile of insects would probably weigh more. Insects are the most abundant animals in the world today.

3

Fossils of insects are rare, but those found in amber are perfectly preserved. Even the delicate veins in the wings as well as the segments of the body, antennae, and legs are visible. Amber is fossil resin, in which the insects were trapped before it hardened.

FOSSILS of primitive, wingless insects are found in rocks 350 million years old. About 300 million years ago, giant ancestors of dragonflies, with wing-spreads of 2½ feet, soared over the coal swamps. They were the largest insects that have ever lived. From fossils we know that some present-day insects, such as cockroaches and termites, are little changed in form or appearance from their ancient ancestors.

Insects belong to the great group of joint-legged animals, the arthropods. Crustaceans (crabs, lobsters, crayfish, shrimps, etc.), centipedes, millipedes, ticks, and spiders are their close relatives.

The many kinds of insects are basically alike. Their body is divided into three parts: head, thorax, and abdomen. Attached to the head is a pair of antennae, or feelers. Most insects have compound eyes. Such eyes are made up of many simple units—as many as 25,000 in some species. Attached to the thorax are three pairs of legs and, in most species, two pairs of wings. Insects breathe through spiracles, openings along the sides of the abdomen. Spiders, which are *not* insects, have only two body divisions, no wings, and four pairs of legs.

PARTS OF A TYPICAL INSECT

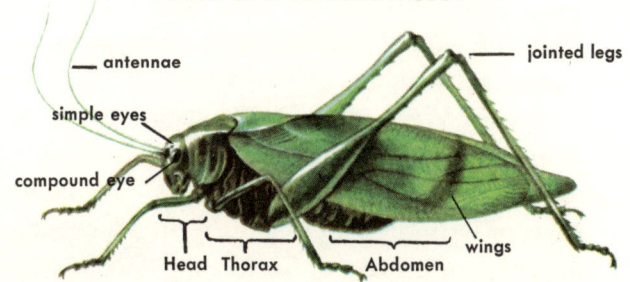

antennae

simple eyes

compound eye

jointed legs

wings

Head Thorax Abdomen

Chewing-lapping (Honeybee)

Piercing-sucking (Mosquito)

Sucking (Butterfly)

Sponging (House Fly)

TYPES OF MOUTHPARTS

HOW INSECTS FEED Grasshoppers, beetles, wasps and ants are among the kinds of insects that bite or crush their food with their heavy jaws. They have chewing mouth-parts. Mosquitoes, bugs and some lice stick their sharp beak, which often has slim, knife-like cutting parts, into a plant or an animal and then suck out the juices or blood. With its sponging mouth-parts, the House Fly can only take in liquids. It dissolves solids with its saliva before it feeds. The mouth-parts of butterflies and moths form a long sucking tube, or proboscis, for feeding on nectar.

HOW INSECTS GROW In primitive, wingless insects the egg hatches into young that are like the adult except in size. Grasshoppers, cockroaches, crickets and true bugs have three life stages—egg, nymph, adult (page 22). Nymphs look like the adults but lack wings. Butterflies and moths, flies, beetles, ants, bees, wasps and others have four stages: egg, larva, pupa, and adult (pp. 27 and 30). Larvae, called caterpillars, maggots, and grubs, do not look like the adults and usually live in a different kind of place. In the resting or pupa stage the larva becomes an adult.

Development of a Bristletail, or Silverfish, a primitive, wingless insect.

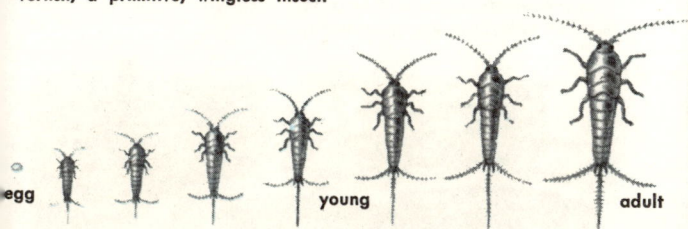

egg   young   adult

GRASSHOPPERS feed on many kinds of plants. A great swarm of grasshoppers—sometimes in sky-darkening clouds covering hundreds of square miles—destroys all vegetation in its path. In Bible days, grasshoppers were called "locusts," a name still used in some places. Locust plagues still cause famines in some parts of the world. On the other hand, people in some countries eat grasshoppers, usually roasted or fried. These insects are also an important food of many kinds of birds and other small animals. More than 10,000 kinds are known, from every part of the world except the polar regions.

The **RED-LEGGED GRASSHOPPER** (1.5 in.), of western U.S., damages many crops. At left, a female is depositing a packet of eggs in the soil. At right, a nymph is emerging from egg.

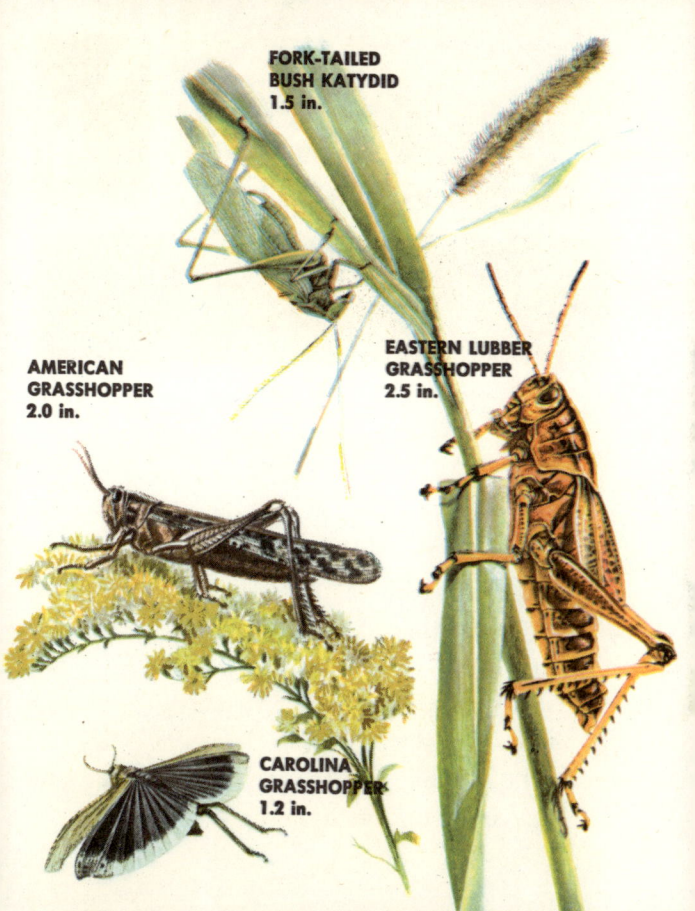

**FORK-TAILED
BUSH KATYDID**
1.5 in.

**EASTERN LUBBER
GRASSHOPPER**
2.5 in.

**AMERICAN
GRASSHOPPER**
2.0 in.

**CAROLINA
GRASSHOPPER**
1.2 in.

FIELD CRICKET 0.9"

eggs

Female **FIELD CRICKET** (0.9 in.) is laying her eggs in the soil. Tree crickets lay eggs in slits cut in stems of plants, often causing that part of the plant to die.

CRICKETS are known throughout the world for their singing. In the Orient, they are kept in cages, like song-birds, and the males are pitted against each other in fights. Black field crickets are found in nearly every country. The more delicate, less commonly seen tree crickets sing the loudest, however. Mole crickets, which have broad, paddle-like front legs, burrow in the soil and are sometimes serious crop pests.

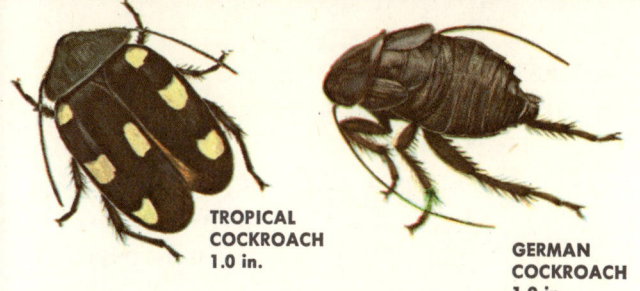

TROPICAL COCKROACH
1.0 in.

GERMAN COCKROACH
1.0 in.

COCKROACHES, like crickets, are related to grasshoppers, walking-sticks and praying mantids. Most of the more than 2,000 kinds are found in the tropics. In temperate regions they are commonly pests in houses. Cockroaches are active at night and hide during the day.

AMERICAN COCKROACH
1.5 in.

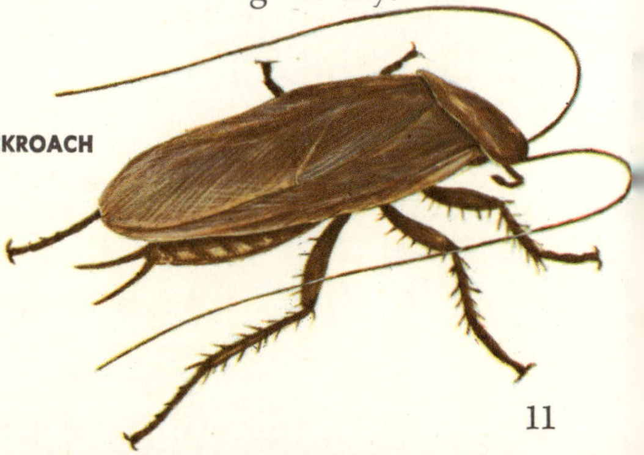

WALKING-STICKS, as their name implies, are slim and stick-like. The most unusual of the 2,000 or so kinds look remarkably like leaves; even the veins in the wings are like the veins in a leaf. Walking-sticks eat plants and in some regions are pests in forests. Walking-sticks are most plentiful and varied in southeast Asia, while the closely related praying mantids are common in South Africa. Both groups occur throughout the world, however. Mantids eat other insects. They use their powerful, sharp-spined front legs to grasp and hold their prey.

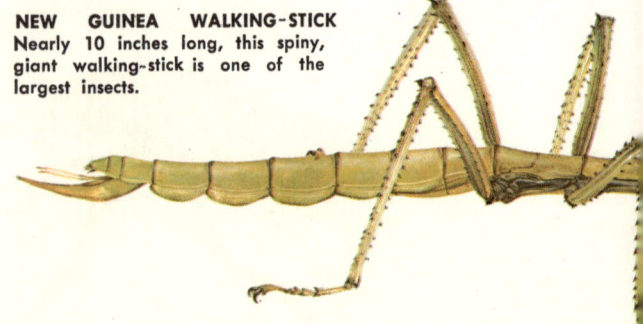

NEW GUINEA WALKING-STICK
Nearly 10 inches long, this spiny, giant walking-stick is one of the largest insects.

ORIENTAL MANTID (3.5 in.) waits for prey to come within reach. Four species are found in the U.S.

13

TERMITES are primitive insects that live in highly organized colonies. Within a colony, different looking termites do different kinds of work. The blind and soft-bodied workers die if they are exposed to the sun, and so the termite nest is kept sealed, except when temporary openings are made to let out winged swarmers. Termites eat dead wood, converting it into soil; thus, in nature they are useful. But they also destroy buildings and other wood structures. Termites cannot digest the wood they eat. One-celled animals that live in the termites' intestines do this. Termites of the tropics build above ground nests, as hard as concrete and as much as 20 feet tall and 50 feet around.

CASTES OF SUBTERRANEAN TERMITE

worker

soldier

queen

14

egg

LICE AND FLEAS Though members of different groups, both of these insects are parasites. Most kinds feed on blood. They may occur on only one kind of host animal and on only one area of that animal's body. Some lice lock their claws around a hair on their host's body. Fleas escape danger by jumping or by scurrying out of sight in the host's hair or fur. Both fleas and lice may transmit diseases, such as typhus (lice) and bubonic plague (fleas).

DOG FLEA
0.1 in.

15

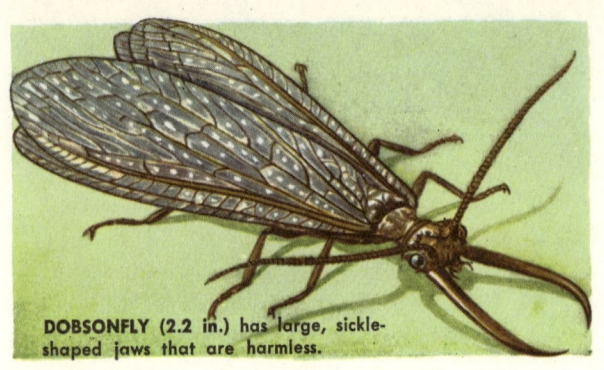

DOBSONFLY (2.2 in.) has large, sickle-shaped jaws that are harmless.

DOBSONFLIES are the winged adults of hellgrammites, the more familiar larvae of the same insect. Large, black, and with powerful jaws, hellgrammites live in streams and feed on small animals. Hellgrammites grow in the water for about three years before changing into adults.

Ant-lions, closely related to Dobsonflies, look like little dragonflies but are much more delicate. The larva hides at the bottom of a funnel-shaped pit dug in the sand. If an ant starts into the pit, the Ant-lion uses its broad shovel-shaped

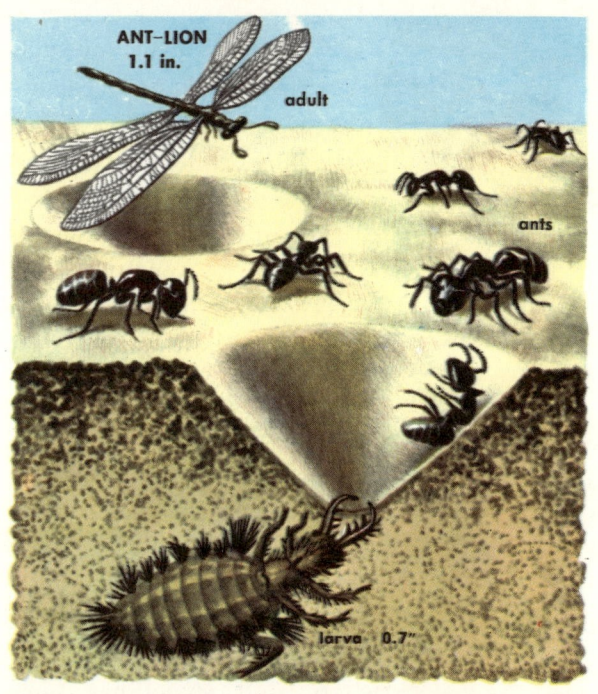

ANT-LION
1.1 in.

adult

ants

larva 0.7"

head to throw sand onto the ant and keep it from
crawling out. When the ant slides to the bottom,
it is grabbed in the Ant-lion's sickle-shaped jaws.

17

POTATO LEAF-HOPPER
0.3 in.

ROSE LEAF-HOPPER
0.3 in.

THREE-BANDED LEAF-HOPPER 0.3 in.

LEAF-HOPPERS, and such related insects as tree hoppers and spittlebugs, suck the juices from plants. Some are serious pests, causing infested plants to wither and die. Many of the thousands of kinds of leaf-hoppers found throughout the world transmit plant diseases. The much larger cicadas, members of the same group, are best known for the mating calls of the males. Their loud, continuous singing is almost deafening. They sing during the day and sometimes long into the night. The largest, found in Borneo, has an 8-inch wingspread.

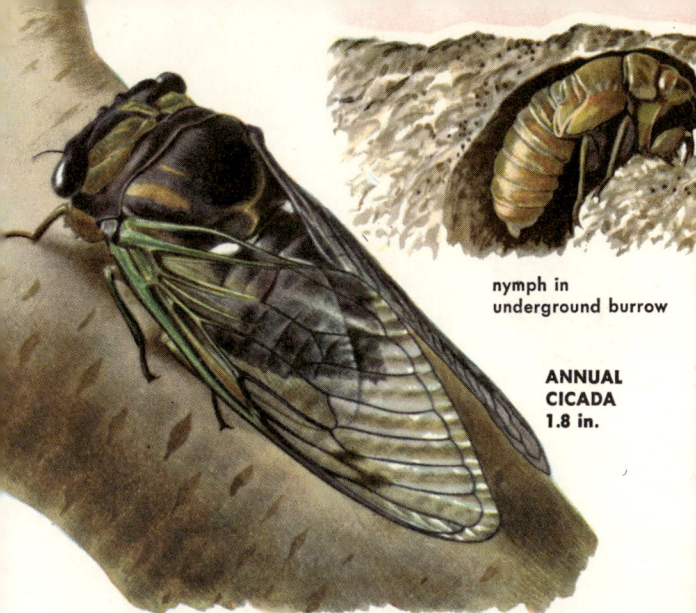

nymph in
underground burrow

**ANNUAL
CICADA
1.8 in.**

The female Annual Cicada lays her eggs in slits
cut in stems or in twigs. The nymphs live underground, feeding on roots. The Periodical Cicada,
which emerges in 13- and 17-year cycles, has red
eyes; an attractive Chinese species has a scarlet
and black body.

OYSTER–SHELL SCALE 0.1 in.

SAN JOSE SCALE 0.1 in.

SCALES are tiny insects belonging to the same group as leaf-hoppers and cicadas. They live, in large numbers, on twigs and leaves, sucking out the sap. Many kinds are pests in orchards. Nymphs, or "crawlers," are active. Adult females secrete a waxy or powdery covering and do not move; the males are winged. Shellac is obtained from the secretions of a scale insect of southeastern Asia.

BUGS belong to the order Hemiptera, which means "half wing." The base of the front wings is leathery; the rear portion is a thin membrane. Bugs have piercing or sucking mouth-parts. Some feed on the sap of plants, others on the blood of animals. Many species are agricultural pests. Bugs may harm plants by injecting poisons as they feed, and those that feed on animals may transmit disease. Also, their bites may be painful. Some kinds of bugs are valued by man, as they feed on other insects that are pests. The word "bug" is commonly used to refer to any insect but is used correctly only for this group of insects.

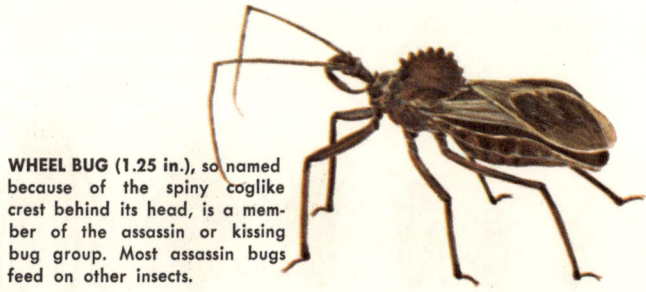

WHEEL BUG (1.25 in.), so named because of the spiny coglike crest behind its head, is a member of the assassin or kissing bug group. Most assassin bugs feed on other insects.

HARLEQUIN BUG
0.4 in.

Stink bugs are among the most familiar of true bugs. This is a large family with members found throughout the world. All are plant feeders. Most stink bugs of cooler regions are grey, brown, or green, but the Harlequin Bug, a serious pest of cabbage and related plants, is a colourful black and red. Many stink bugs of the tropics are purple, scarlet, orange, blue, or black and have odd-shaped swellings on their legs and antennae. Squash bugs, chinch bugs and tiny lace bugs are also common plant pests.

CHINCH BUG nymphs, which hatch from eggs, look much like adults but lack wings. Each time they grow and shed, the nymphs become larger and their wings develop more, until they are adults.

egg

nymphs

adult
0.2 in.

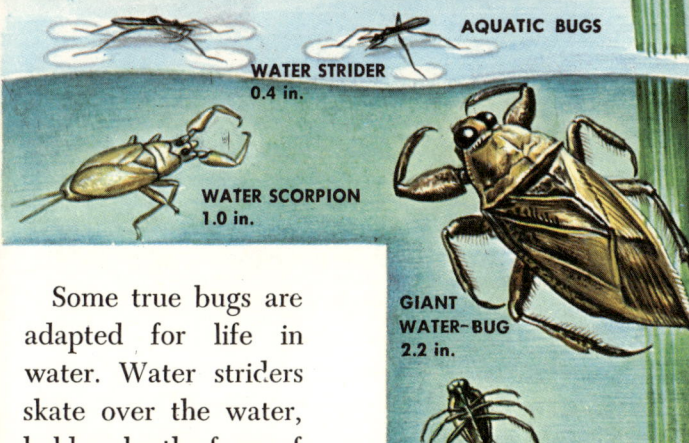

AQUATIC BUGS

WATER STRIDER
0.4 in.

WATER SCORPION
1.0 in.

GIANT
WATER-BUG
2.2 in.

BACK-SWIMMER
0.5 in.

Some true bugs are adapted for life in water. Water striders skate over the water, held up by the force of surface tension. They skim along on the surface film and do not break through. Water scorpions prey on other aquatic insects. They breathe through a pair of air tubes at the end of their abdomen. Giant water-bugs often come out of water and fly to lights. They feed on water insects and even on small fish. One kind from Brazil is more than 4 inches long. Back-swimmers, like giant water-bugs, can bite severely.

DRAGON-FLIES and the closely related damsel-flies are among the few groups of insects that live in water in their immature stages. Only about 3 per cent of all insects belong to these groups. The nymphs of dragon-flies, called naiads, do not resemble the adults in appearance or in habits. Naiads live in fresh water and eat insects. In turn they are an important food of fish and other animals. Dragon-flies are swift fliers, skilful in hovering. A South American species has a wing-spread of nearly 8 inches, while tiny damselflies of Australia have wing-spreads of less than an inch. Dragon-flies rest with their wings stretched out. Damselflies fold their wings and hold them over their body when resting. Neither dragon-flies nor damsel-flies can sting.

GREEN DARNER
2.6 in.

nymph, or
naiad

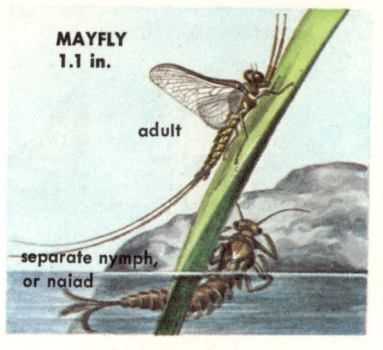

MAYFLY
1.1 in.

adult

separate nymph,
or naiad

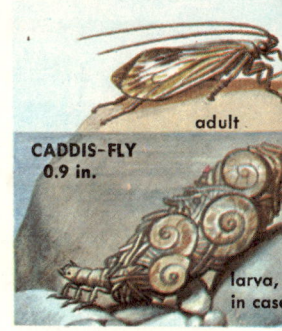

adult

CADDIS-FLY
0.9 in.

larva,
in case

MAYFLIES often live only a few hours as adults, only long enough to mate and to lay eggs. Sometimes millions of adults emerge at the same time. They are fragile and are poor fliers. Mayfly naiads may develop for as long as three years.

CADDIS-FLY larvae live in silk-lined cases covered with twigs, pebbles, shells, or other debris. Only their head and forepart of their body stick out of the case. Though most eat plants, some larvae capture small water animals in funnel-shaped nets. Adults are moth-like.

Most moths have feathery antennae, while those of butterflies are knobbed.

At rest, a moth's wings are outstretched; a butterfly's are folded.

BUTTERFLIES AND MOTHS, because of the beauty of their scaly wings, are perhaps the best known of all insects. Butterflies are largely active in the day, moths at night. Moths are much more common, and many species, in the larval or caterpillar stage, damage crops and stored products. Adults feed on nectar. The domestic Silkworm Moth has been raised in captivity for so many centuries that it can no longer survive in the wild. A few kinds of butterflies, such as the Monarch (p. 27) and the Painted Lady, are migratory. They travel south as winter approaches and return in the spring.

Many moths and butterflies produce more than one brood a year, though the life cycles of a few kinds take as long as four years. Each passes through four stages: egg, larva (caterpillar), pupa or resting stage, and adult. Many moths pass the pupa stage in a silken cocoon. Other moths and most butterflies pass through the pupa stage in a protective case in the soil or attached to an object above ground.

LIFE CYCLE OF MONARCH BUTTERFLY

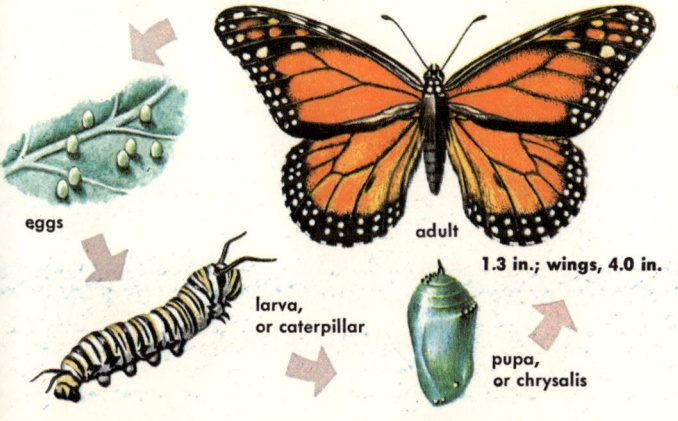

eggs

adult

1.3 in.; wings, 4.0 in.

larva, or caterpillar

pupa, or chrysalis

TIGER SWALLOW-TAIL
0.9 in.; wings, 4 in.

One of the most beautiful of all butterflies is a swallowtail of Asia. It has a wing-spread of 5½ inches. The Giant Swallow-tail is the largest in North America, with a 4½ inch wing-spread. The Tiger Swallow-tail has a wing-spread of nearly 4 inches. Morpho butterflies of tropical America are sought by many collectors. Their shiny metallic, blue and green wings are used to make art objects and jewelry.

**WHITE-LINED
SPHINX MOTH**
1.2 in.; wings, 3.1 in.

WESTERN PYGMY BLUE
0.2 in.; wings, 0.6 in.

Moths are even more varied in size and habits than butterflies. The largest, the giant Atlas Moth of southern Asia, has a wing-spread of nearly 11 inches. The Cecropia Moth is the largest moth in North America.

CECROPIA MOTH
1.2 in.; wings, 5.5 in.

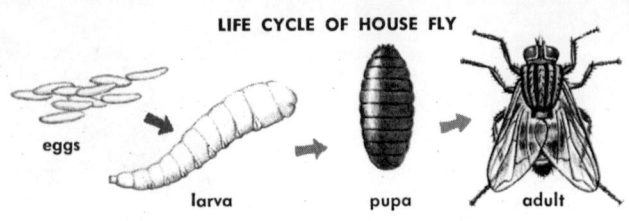

LIFE CYCLE OF HOUSE FLY

eggs

larva

pupa

adult

The House Fly is a carrier of several diseases. Note that it has the same four stages of development as butterflies.

FLIES have only one pair of wings, while most other insects have two. Their hind wings have become a pair of knob-like structures, called halteres, that serve as balancing organs in flight. Some flies, such as mosquitoes, are carriers of disease; others bite or torment. Most of the nearly 100,000 different kinds of flies are harmless, and many are beneficial. The maggots (larvae) of some flies aid in decay of animal and vegetable matter. Other flies prey on insect pests, and some pollinate flowers. Some of the tiny midges are only 1/16 of an inch long, while an **Australian crane** fly has a wing-spread of nearly 4 inches.

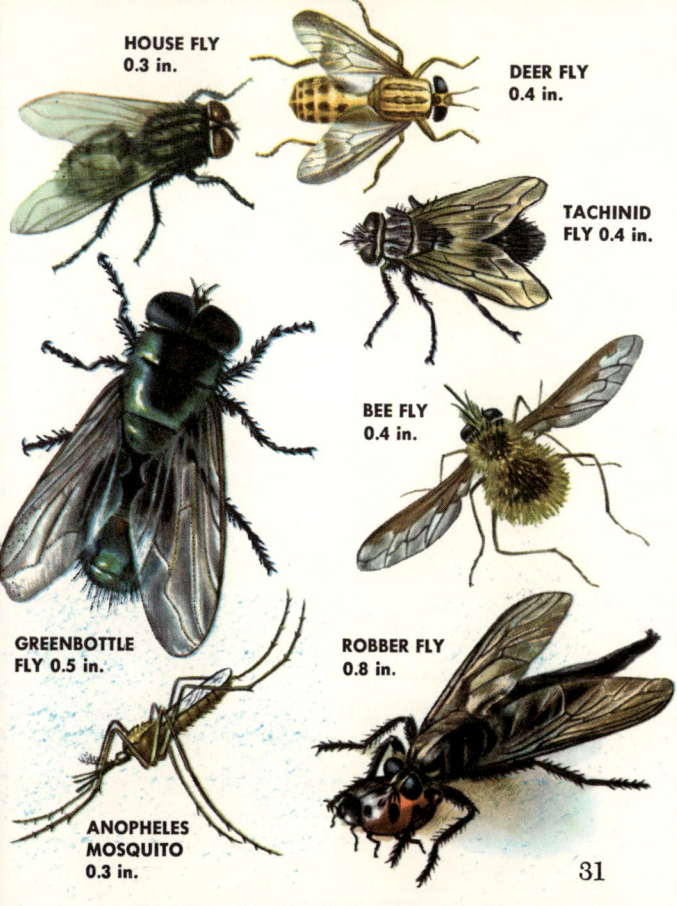

HOUSE FLY
0.3 in.

DEER FLY
0.4 in.

TACHINID
FLY 0.4 in.

BEE FLY
0.4 in.

GREENBOTTLE
FLY 0.5 in.

ROBBER FLY
0.8 in.

ANOPHELES
MOSQUITO
0.3 in.

31

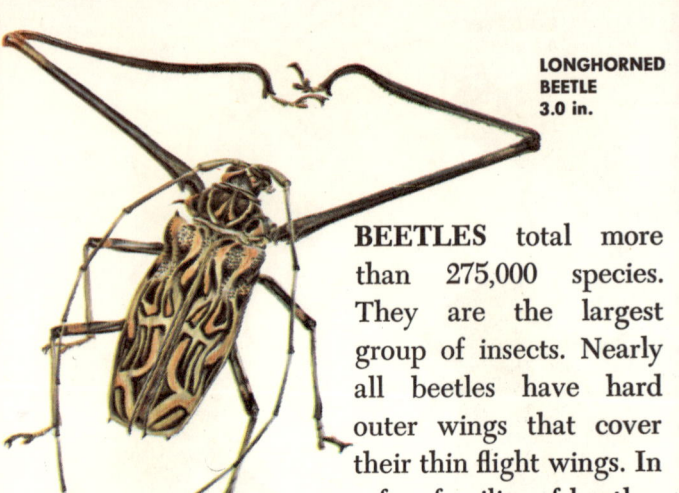

BEETLES total more than 275,000 species. They are the largest group of insects. Nearly all beetles have hard outer wings that cover their thin flight wings. In a few families of beetles, the front wings are soft or are short. Male fireflies have soft front wings, and the females are wingless. Beetle larvae are commonly called grubs. Tiny beetles that feed on fungi are scarcely visible to the naked eye. The giant African Goliath Beetle, 5 inches long and 2 inches wide, is the bulkiest of all insects.

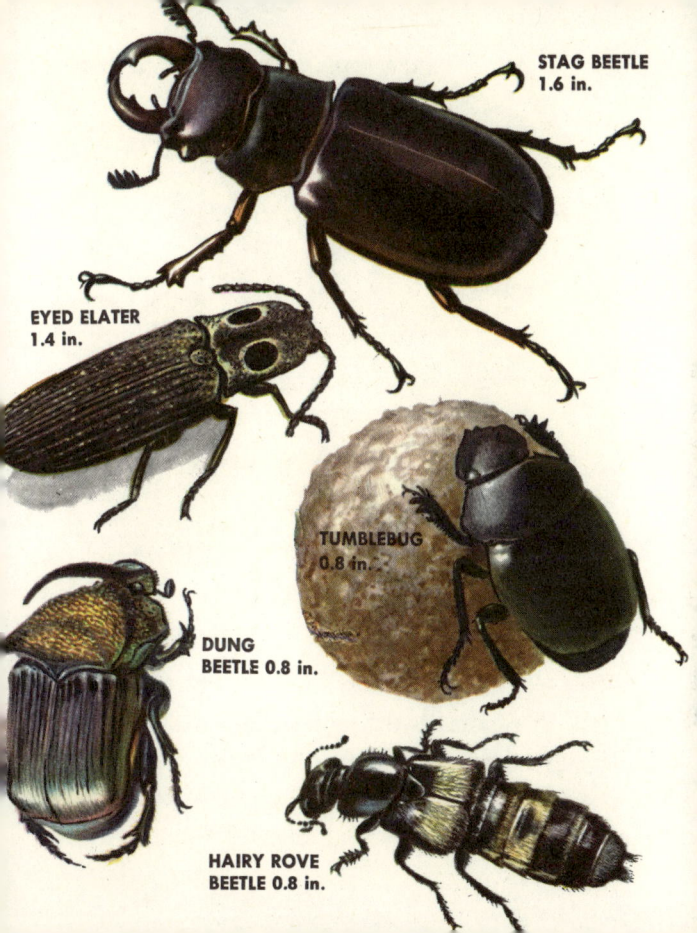

STAG BEETLE
1.6 in.

EYED ELATER
1.4 in.

TUMBLEBUG
0.8 in.

DUNG
BEETLE 0.8 in.

HAIRY ROVE
BEETLE 0.8 in.

MEXICAN BEAN BEETLE
0.3 in.

larva

eggs

Beetles have chewing mouth-parts. Many kinds feed on plants, and among them are some that damage crops. Both the adults and the larvae may be pests, as in the case of the Mexican Bean Beetle above. Ladybird beetles belong to the

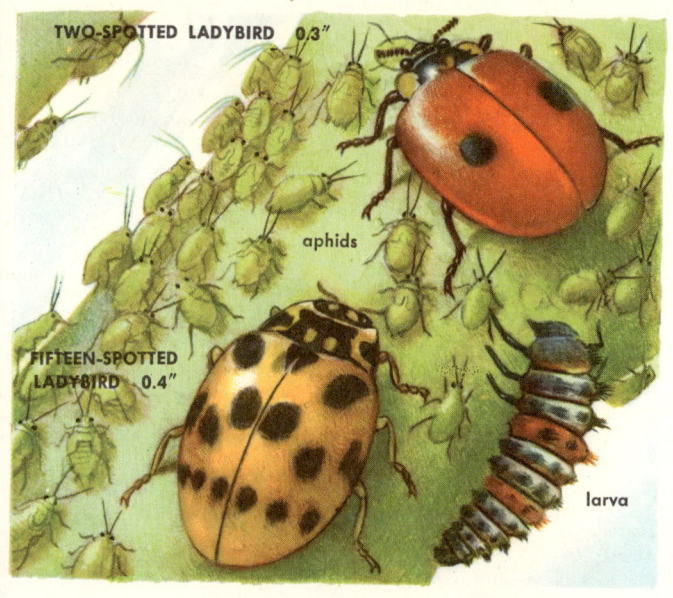

TWO-SPOTTED LADYBIRD 0.3"

aphids

FIFTEEN-SPOTTED LADYBIRD 0.4"

larva

same family but do not eat plants. Both the adults and larvae of ladybird beetles prey on aphids—insects which suck the sap from plants. Ladybird beetles are released in large numbers in orchards to control aphids and scales.

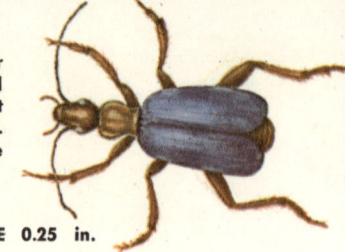

When pursued, a Bombardier Beetle turns up its tail and ejects a spray of liquid that appears like a puff of smoke. The stinging gas gives the beetle time to escape.

BOMBARDIER BEETLE 0.25 in.

Beetles grow and develop the same way as butterflies and flies. Eggs hatch into larvae or grubs. After a pupal or resting stage the adult emerges. Beetles are interesting because of their great variety of shapes and colours and because of the unusual habits of some species. Some, for example, feed on spices, and others bore into lead cables.

NUT WEEVIL 0.3

Weevils, also called curculios or snout beetles, have a long snout, often curved, with chewing mouth-parts at its tip. There are more than 40,000 kinds of weevils.

Diving beetles are common in lakes and slow streams. They prey on other water insects, or they may attack aquatic animals as large as tadpoles or small fish. The larvae are equally voracious and are commonly called "water tigers." They crawl out on land to pass the pupa stage in an earthen cell. Adults often collect around lights.

DIVING BEETLE
1.1 in.

Mayfly naiad

larva of Diving Beetle

BUMBLE-BEE

BEES belong to the same order of insects as wasps and ants. Bumble-bees pollinate red clovers and other deep-throated blossoms that the short-tongued honey-bees do not visit. Honey-bees and bumble-bees have three castes or work divisions: workers, drones (males), and queens. The bumble-bees nest in holes and store their honey in wax "pots."

Honey-bees have been domesticated for centuries. They produce honey and beeswax worth millions of dollars every year but are even more valuable for another reason. Bees carry the pollen from flower to flower; without this no fruits would form. New colonies of honey-bees form when the old queen flies away and takes part of the workers with her. This is called swarming. A colony consists of about 50,000 bees, most of them workers. The fat-bodied males, called drones, do no work and are driven from the colony after the queen's mating flight.

CASTES OF THE HONEY-BEE

queen

worker

drone

39

PAPER WASP
0.8 in.

OAK GALL WASP
0.2 in.

WASPS are closely related to bees, but most wasps have slimmer bodies than bees. Also, wasps do not have plumed, or branched, hairs on their body; bees do. Most wasps feed their young spiders or insects, while bees feed their larvae honey and pollen. Those wasps that do collect pollen and nectar do not have pollen baskets on their legs, as bees do. Like bees, most wasps sting when disturbed. The sting is painful and may even be dangerous.

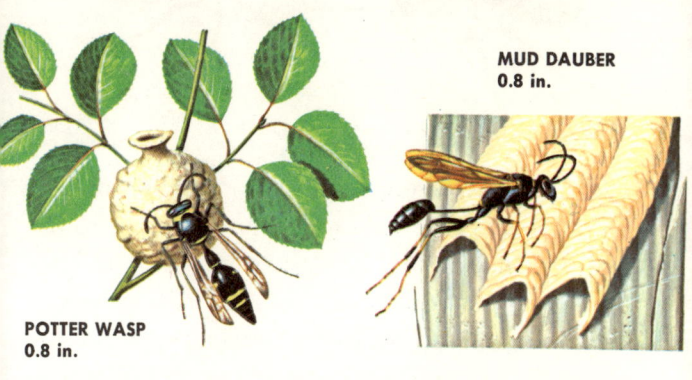

MUD DAUBER
0.8 in.

POTTER WASP
0.8 in.

Same wasp larvae tunnel in plant stems and thus are pests. Others pollinate flowers. The flower of the fig tree can be fertilized only by a small wasp. Other wasps cause swellings, called galls, on leaves and stems in which the females lay their eggs. Nest-building wasps such as hornets, make large nests of chewed-up wood, which house the entire colony.

The tiniest insects known are wee wasps that lay their eggs inside the eggs of other insects.

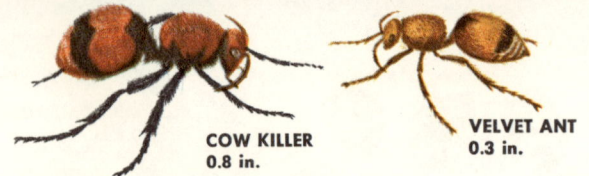

COW KILLER
0.8 in.

VELVET ANT
0.3 in.

These wingless female wasps lay eggs in nests of other wasps.

Some of these wasps are important in controlling insect pests and have been imported to the United States from foreign countries. The largest wasp is the female of a North American ichneumon; she measures 4½ inches from her head to the tip of her egg-laying tubes. She uses her long egg-laying tubes, or ovipositors, to deposit her eggs in the burrow of a wood-boring sawfly, another type of wasp. The ichneumon illustrated here is one of some 3,000 species in North America.

ICHNEUMON WASP
1.4 in.

CICADA KILLER
1.5 in.

The Cicada Killer's sting paralyses but does not kill the cicada, which the wasp drags into its burrow. When the Cicada Killer's egg hatches, the cicada will be food for the larva.

Though small compared to its intended victim, the Tarantula Hawk nearly always wins its battle. It paralyses the large spider by stinging it repeatedly, then puts the spider in a burrow and lays an egg on it. The wasp's larva feeds on the spider.

TARANTULA

TARANTULA HAWK
1.25 in.

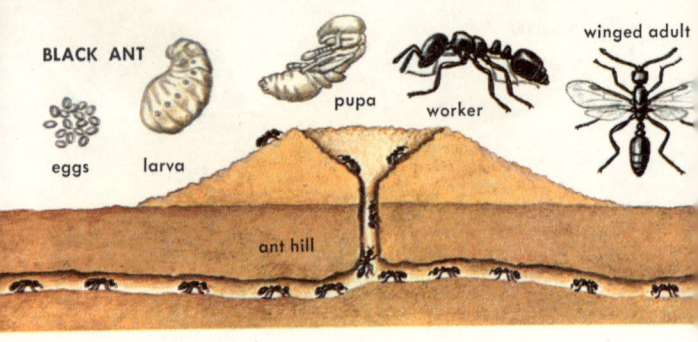

BLACK ANT

eggs   larva   pupa   worker   winged adult

ant hill

Ants live in organised colonies, like termites and bees.

ANTS are said to outnumber all other land-dwelling animals. The more than 6,000 species are found from arctic regions to equatorial rain forests and from deserts to seashores. Large colonies may contain as many as half a million ants. Ants can be told from wasps and bees, to which they are related, by their crooked or elbowed antennae and by the wedge-shaped bump on the slim connection between their abdomen and thorax. At times winged ants appear and fly away to start new colonies.

Winged ants are sometimes mistaken for the winged stage of termites. But ants have a stem-like waist; in termites, the connection is broad.

Most ants nest in the ground. Cones of sand around the entrance holes identify the common ants found in lawns. A woodland ant of Europe piles up mounds of leaves and twigs three or four feet high. The large sun-baked mounds of the Fire Ant damage farm equipment.

Ants will move their nest to a new location if the old one becomes unsatisfactory. In summer, some ants nest in the open in fields or meadows. In winter they make a nest deeper in the ground.

CARPENTER ANT
0.5 in.

FIRE ANT
0.2 in.

CORNFIELD ANT
0.2 in.

HONEYPOT ANT
0.2 in.

ARMY ANT
0.2 in.

LEAFCUTTER, PARASOL, OR ATTA ANT

Parasol or Atta Ants of tropical America cut pieces from the leaves of trees and carry them to their nest. A fungus, grown on the decaying leaves, is eaten by the adults and is fed also to the larvae as they develop.

QUIZ-ME

Here are some questions you can answer if you have studied this book. The pages where the answers will be found are listed at the end.

1 What insect has been domesticated for so many years that it can no longer survive in the wild?
2 What insects may change the location of their nests with the season?
3 What are hellgrammites?
4 What crop is pollinated by bumble-bees and not by honey-bees?
5 The young of true bugs go by what name?
6 What are the stages in the life cycles of a butterfly?
7 What are the tiniest insects known?
8 What were the largest insects that have ever lived?
9 What is the largest butterfly in North America?
10 What is another name for the larvae of flies?
11 What insect songsters are kept in cages?
12 Of what material is a hornet's nest made?
13 Where do cicadas lay their eggs?
14 About how many species of insects have been named?

15 The females of what group of wasps are wingless?
16 What insects were the "locusts" referred to in the Bible?
17 From what kind of insect is shellac obtained?
18 Which insects are said to outnumber in individuals all other land-dwelling animals combined?
19 What is the largest moth in North America?
20 Which insects may live as adults for only a few hours?
21 What is the largest group of insects?
22 Why are ladybird beetles helpful?
23 Which insects have only two wings rather than four?
24 Why do termites keep their nests sealed from air?

ANSWERS: 1 (p. 26), **2** (p. 45), **3** (p. 16), **4** (p. 38), **5** (p. 22), **6** (p. 27), **7** (p. 41), **8** (p. 4), **9** (p. 28), **10** (p. 30), **11** (p. 10), **12** (p. 41), **13** (p. 19), **14** (p. 3), **15** (p. 42), **16** (p. 8), **17** (p. 20), **18** (p. 44), **19** (p. 29), **20** (p. 25), **21** (p. 32), **22** (p. 35), **23** (p. 30), **24** (p. 14).

ILLUSTRATIONS BY: *Arch and Miriam Hurford, James Gordon Irving, Enid Kotschnig, Joe Lombardaro, John Wood*
COVER: *John Wood*

B